Editor
Gisela Lee

Editorial Manager
Karen Goldfluss, M.S. Ed.

Editor-in-Chief
Sharon Coan, M.S. Ed.

Cover Artist
Barb Lorseyedi

Art Coordinator
Kevin Barnes

Art Director
CJae Froshay

Imaging
Alfred Lau
James Edward Grace
Temo Parra

Product Manager
Phil Garcia

Publisher
Mary D. Smith, M.S. Ed.

Authors

Teacher Created Resources Staff

Teacher Created Resources, Inc.
12621 Western Avenue
Garden Grove, CA 92841
www.teachercreated.com
ISBN: 978-0-7439-3317-9
©*2002 Teacher Created Resources, Inc.*
Reprinted, 2019
Made in U.S.A.

Table of Contents

Introduction

The old adage "practice makes perfect" can really hold true for your child and his or her education. The more practice and exposure your child has with concepts being taught in school, the more success he or she is likely to find. For many parents, knowing how to help their children can be frustrating because the resources may not be readily available. As a parent it is also difficult to know where to focus your efforts so that the extra practice your child receives at home supports what he or she is learning in school.

This book has been designed to help parents and teachers reinforce basic skills with your children. *Practice Makes Perfect* reviews basic math skills for children in the first and second grades. The math focus is on telling time and concepts related to it. While it would be impossible to include all concepts taught in the first and second grades in this book, the following basic objectives are reinforced through practice exercises. These objectives support math standards established on a district, state, or national level. (Refer to the Table of Contents for the specific objectives of each practice page.)

- identifying the hour and minute hands on a clock
- word problems
- elapsed time
- understanding graphs, charts, and tables with relation to time

- telling time to the minute, quarter 'til, and quarter past the hour, half hour, and hour
- understanding days, dates, and calendars with relation to time
- differentiating between A.M. and P.M.

There are 36 practice pages organized sequentially, so children can build their knowledge from more basic skills to higher-level math skills. To correct the practice pages in this book, use the answer key provided on pages 47 and 48. Six practice tests follow the practice pages. These provide children with multiple-choice test items to help prepare them for standardized tests administered in schools. As children complete a problem, they fill in the correct letter among the answer choices. An optional "bubble-in" answer sheet has also been provided on page 46. This answer sheet is similar to those found on standardized tests. As your child completes each test, he or she can fill in the correct bubbles on the answer sheet.

How to Make the Most of This Book

Here are some useful ideas for optimizing the practice pages in this book:

- Set aside a specific place in your home to work on the practice pages. Keep it neat and tidy with materials on hand.

- Set up a certain time of day to work on the practice pages. This will establish consistency. An alternative is to look for times in your day or week that are less hectic and conducive to practicing skills.

- Keep all practice sessions with your child positive and constructive. If the mood becomes tense, or you and your child are frustrated, set the book aside and look for another time to practice.

- Help with instructions if necessary. If your child is having difficulty understanding what to do or how to get started, work through the first problem with him or her.

- Review the work your child has done. This serves as reinforcement and provides further practice.

- Allow your child to use whatever writing instruments he or she prefers. For example, using colored pencils can add variety and pleasure to drill work.

- Pay attention to the areas in which your child has the most difficulty. Provide extra guidance and exercises in those areas. Allowing children to use drawings and manipulatives, such as coins, tiles, game markers, or flash cards, can help them grasp difficult concepts more easily.

- Look for ways to make real-life applications to the skills being reinforced.

Practice 1

Directions: Write the times in the boxes below at the bottom of each clock. Use the times in the boxes at the bottom of this page.

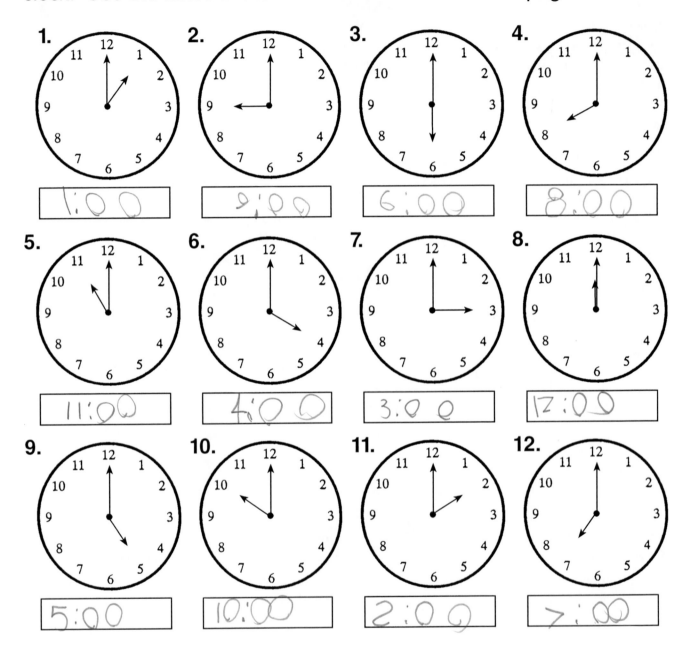

1.

1:00

2.

9:00

3.

6:00

4.

8:00

5.

11:00

6.

4:00

7.

3:00

8.

12:00

9.

5:00

10.

10:00

11.

2:00

12.

7:00

7:00	4:00	6:00	5:00
10:00	9:00	8:00	1:00
12:00	2:00	11:00	3:00

Practice 2

1. Which clock shows the *hour* hand pointing to 6?

(A)

(B)

(C)

(D)

2. Which clock shows the *hour* hand pointing to 10?

(A)

(B)

(C)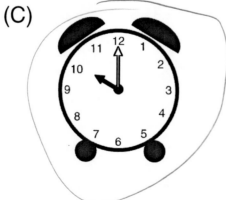

(D)

Practice 3

1. Which clock shows 8:00?

(A)

(B)

(C)

(D)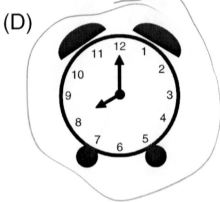

2. Which clock shows 3:00?

(A)

(B)

(C)

(D)

Practice 4

1. Which clock shows 2:00?

(A)

(B)

(C)
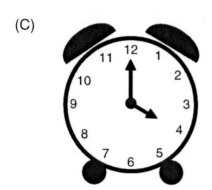

(D)
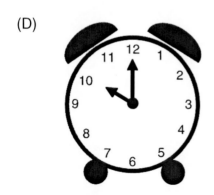

2. Which clock shows 4:00?

(A)

(B)
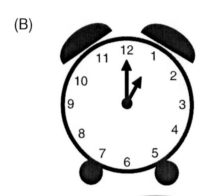

(C)
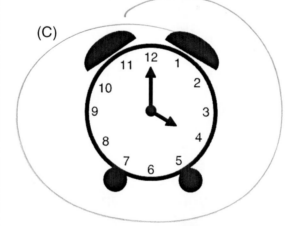

(D)
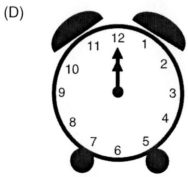

Practice 5

Write the time.

___10:00___ o'clock

Write the time.

___5:00___ o'clock

Write the time.

___7:00___ o'clock

Draw the hands on the clock to show the time that you get up each day.

Draw the hands on the clock to show the time that you go to sleep.

Draw both hands.

6 o'clock

Draw both hands.

9 o'clock

Draw both hands.

3 o'clock

Draw the hands on the clock to show the time that you go to school.

Draw the hands on the clock to show the time that you eat lunch.

Practice 6

1.

What time is it?

5:00

2.

What time is it?

_____

3.

What time is it?

_____

4.

What time is it?

_____

5.

What time is it?

_____

6.

What time is it?

_____

Practice 7

1.

What time is it?

_____

4.

What time is it?

_____

2.

What time is it?

_____

5.

What time is it?

_____

3.

What time is it?

_____

6.

What time is it?

_____

Practice 8

1. Which clock shows the *minute* hand pointing to 1?

(A)

(B)

(C)

(D)

2. Which clock shows the *minute* hand pointing to 2?

(A)

(B)

(C)

(D)

Practice 9

1. Which clock shows the *minute* hand pointing to 11?

(A)

(B)

(C)

(D)

---

2. Which clock shows the *hour* hand pointing to 9?

(A)

(B)

(C)

(D)

Practice 10

1.

The *minute* hand is

pointing to _____.

4.

The *hour* hand is

pointing to _____.

2.

The *hour* hand is

pointing to _____.

5.

The *minute* hand is

pointing to _____.

3.

The *minute* hand is

pointing to _____.

6.

The *minute* hand is

pointing to _____.

Practice 11

1.

The *hour* hand is

pointing to _____ .

2.

The *hour* hand is

pointing to _____ .

3.

The *minute* hand is

pointing to _____ .

4.

The *minute* hand is

pointing to _____ .

5.

The *minute* hand is

pointing to _____ .

6.

The *minute* hand is

pointing to _____ .

Practice 12

1.

What time is it?

_____

4.

What time is it?

_____

2.

What time is it?

_____

5.

What time is it?

_____

3.

What time is it?

_____

6.

What time is it?

_____

Practice 13

1. What time does the clock show?

(A) 4:30   (B) 4:00

(C) 3:30   (D) 3:00

3. What time does the clock show?

(A) 10:00   (B) 11:00

(C) 11:30   (D) 10:30

2. What time does the clock show?

(A) 10:00   (B) 9:30

(C) 10:30   (D) 9:00

4. What time does the clock show?

(A) 1:00   (B) 2:30

(C) 1:30   (D) 2:00

Practice 14

1. What time does the clock show?

(A) 8:00   (B) 7:30

(C) 8:30   (D) 7:00

3. What time does the clock show?

(A) 12:30   (B) 1:30

(C) 1:00   (D) 12:00

2. What time does the clock show?

(A) 6:00   (B) 5:30

(C) 5:00   (D) 6:30

4. What time does the clock show?

(A) 9:30   (B) 7:30

(C) 10:30   (D) 8:30

Practice 15

1.

Draw the hands on the clock to show 11:30.

4.

Draw the hands on the clock to show 10:30.

2.

Draw the hands on the clock to show 4:30.

5.

Draw the hands on the clock to show 3:30.

3.

Draw the hands on the clock to show 1:30.

6.

Draw the hands on the clock to show 9:30.

Practice 16

1. What time is it?

(A) 2:06   (B) 2:45

(C) 6:02   (D) 6:10

3. What time is it?

(A) 6:35   (B) 6:07

(C) 7:15   (D) 7:06

2. What time is it?

(A) 11:06   (B) 11:15

(C) 6:11   (D) 6:55

4. What time is it?

(A) 6:25   (B) 5:06

(C) 5:15   (D) 6:05

Practice 17

1. What time is it?

(A) 6:05   (B) 5:45

(C) 5:06   (D) 6:25

3. What time is it?

(A) 6:40   (B) 8:06

(C) 6:08   (D) 8:15

2. What time is it?

(A) 1:15   (B) 6:01

(C) 1:06   (D) 6:05

4. What time is it?

(A) 6:10   (B) 10:45

(C) 6:50   (D) 10:06

Practice 18

1. What time is it?

  (A) 6:12   (B) 12:15
  (C) 6:00   (D) 12:06

3. What time is it?

  (A) 6:02   (B) 2:06
  (C) 6:10   (D) 2:15

2. What time is it?

  (A) 6:07   (B) 7:06
  (C) 6:35   (D) 7:45

4. What time is it?

  (A) 11:45   (B) 6:11
  (C) 11:06   (D) 6:55

Practice 19

1. What time is it?

(A) a quarter past 10

(B) a quarter past 9

(C) a quarter to 10

(D) a quarter to 9

3. What time is it?

(A) a quarter to 9

(B) a quarter past 8

(C) a quarter to 8

(D) a quarter past 9

2. What time is it?

(A) a quarter past 11

(B) a quarter to 11

(C) a quarter to 12

(D) a quarter past 12

4. What time is it?

(A) a quarter past 2

(B) a quarter past 1

(C) a quarter to 2

(D) a quarter to 1

Practice 20

1. What time is it?

(A) a quarter to 12

(B) a quarter to 1

(C) a quarter past 1

(D) a quarter past 12

3. What time is it?

(A) a quarter to 7

(B) a quarter to 6

(C) a quarter past 7

(D) a quarter past 6

2. What time is it?

(A) a quarter to 4

(B) a quarter past 4

(C) a quarter past 5

(D) a quarter to 5

4. What time is it?

(A) a quarter to 8

(B) a quarter past 8

(C) a quarter past 9

(D) a quarter to 9

Practice 21

1. Draw hands on the clock to show 5:45.

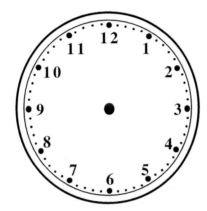

2. Draw hands on the clock to show 8:45.

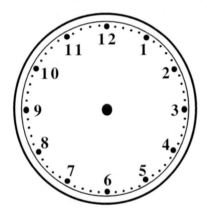

3. Draw hands on the clock to show 7:15.

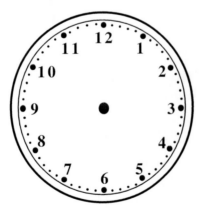

4. Draw hands on the clock to show 6:15.

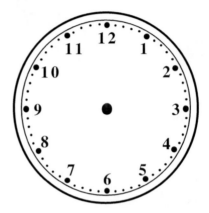

5. Draw hands on the clock to show 8:15.

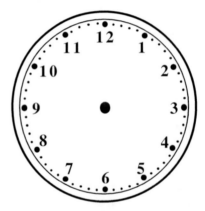

6. Draw hands on the clock to show 11:15.

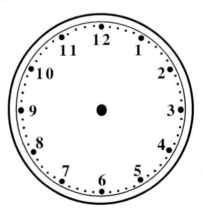

Practice 22

1. Draw hands on the clock to show 5:15.

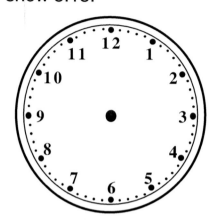

4. Draw hands on the clock to show 9:45.

2. Draw hands on the clock to show 10:15.

5. Draw hands on the clock to show 10:45.

3. Draw hands on the clock to show 11:45.

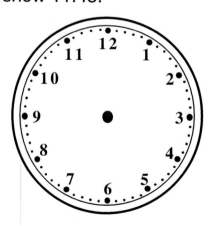

6. Draw hands on the clock to show 9:15.

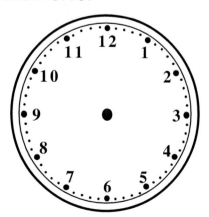

Practice 23

Directions: There are 60 minutes in one hour. The numbers on the face of a clock not only tell the hour, but when counted by 5's, the numbers also tell the number of minutes! Count by 5's to fill in the missing minutes. Write the time on the lines below each clock.

1.

___ : ___ ___

2.

___ : ___ ___

3.

___ : ___ ___

4.

___ : ___ ___

5.

___ : ___ ___

6.

___ : ___ ___

Practice 24

Directions: There are 60 minutes in one hour. The numbers on the face of a clock not only tell the hour, but when counted by 5's, the numbers also tell the number of minutes! Count by 5's to fill in the missing minutes. Write the time on the lines below each clock.

1.

____ : ____ ____

2.

____ : ____ ____

3.

____ : ____ ____

4.

____ : ____ ____

5.

____ : ____ ____

6.

____ : ____ ____

Practice 25

1. What time is it? _____

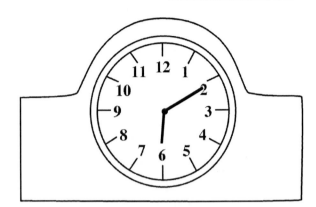

2. What time is it? _____

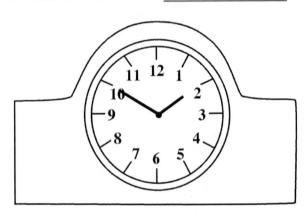

3. What time is it? _____

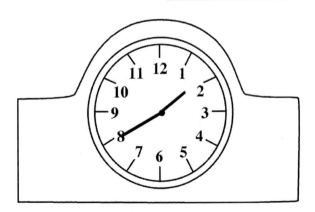

4. What time is it? _____

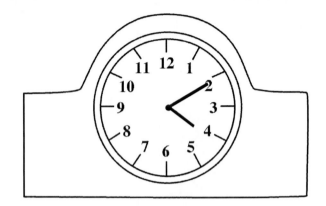

5. What time is it? _____

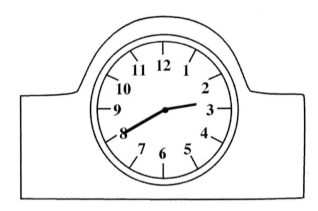

6. What time is it? _____

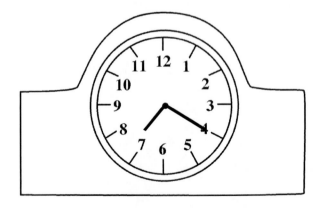

Practice 26

1. What time is it? _____

4. What time is it? _____

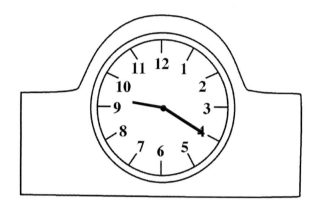

2. What time is it? _____

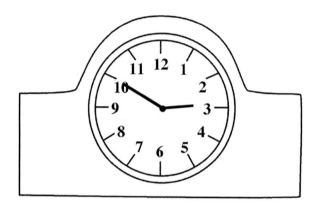

5. What time is it? _____

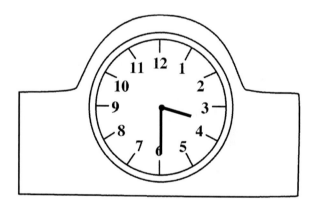

3. What time is it? _____

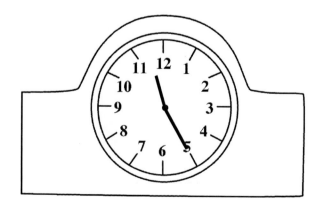

6. What time is it? _____

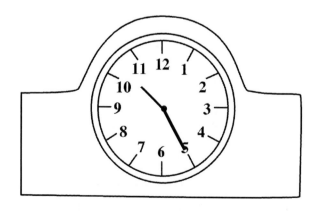

Practice 27

Arthur has a busy afternoon. Help him plan his schedule by filling in the boxes below with the correct times.

1. Arthur's mom picks him up from school at

What time is it?

:

2. His eye doctor appointment is at

What time is it?

:

3. Dr. Iris takes Arthur into her office at

What time is it?

:

4. Arthur and his mother leave Dr. Iris' office at

They arrive at a store thirty minutes later. Show the time.

What time is it?

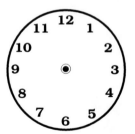

:

:

Practice 28

One type of graph that gives people information is a called a pictograph. In a pictograph pictures are used instead of numbers.

Here is a pictograph that shows the number of used books collected during two-hour time periods for a one-day classroom used-book drive.

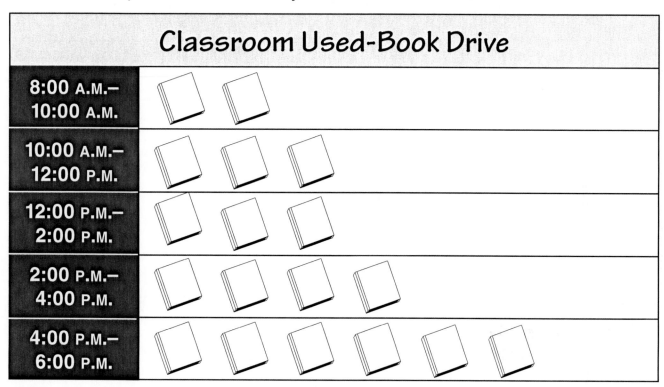

Classroom Used-Book Drive	
8:00 A.M.–10:00 A.M.	📖 📖
10:00 A.M.–12:00 P.M.	📖 📖 📖
12:00 P.M.–2:00 P.M.	📖 📖 📖
2:00 P.M.–4:00 P.M.	📖 📖 📖 📖
4:00 P.M.–6:00 P.M.	📖 📖 📖 📖 📖 📖

Key: 📖 = 5 books

1. Between what hours were the most books collected? _____

2. What was the total number of books collected between 10:00 A.M. and 12:00 P.M.? _____

3. Between what hours were the fewest books collected?

   _____

4. During which times of the day were the same number of books collected?

   _____   _____

5. How many books were collected between the hours of 8:00 A.M. and 12:00 P.M.? _____

Practice 29

Directions: Read and answer each question.

1. Nathan does homework for 15 minutes. If he starts after dinner at 6:30, when does he finish?

   _____

2. Kaylee rides her bike for half an hour. If she stops at 12:45 P.M., when did she begin?

   _____

3. Zack reads for 15 minutes in the morning, 40 minutes in the afternoon, and 20 minutes before bed. How much total time did he spend reading?

   _____

4. Jessica's mom drops her off at her friend Molly's house at 1 o'clock in the afternoon. Her mom promises to pick her up in 3 hours. What time should Jessica's mom return?

   _____

5. Eric makes a batch of chocolate chip cookies. He puts them into the oven at 2:15 P.M. If the recipe says to bake them for 15 minutes, when should he take them out?

   _____

6. Dan's favorite television show lasts half an hour. If the show starts at 7:30 P.M., when is it over?

   _____

7. Tara goes to daycare after school. If she arrives at 3:30 and her dad picks her up at 6:00, how long is she at daycare?

   _____

Practice 30

1. Use A.M. or P.M. to show the correct time of day.

Laura goes to bed at 8:00 _____ each night.

2. Use A.M. or P.M. to show the correct time of day.

Peter eats lunch at 11:30 _____ each morning.

3. Use A.M. or P.M. to show the correct time of day.

Ozzie Owl wakes up at 3:00 _____ each morning.

4. Use A.M. or P.M. to show the correct time of day.

Henrietta Hippo takes her bath each night at 7:00 _____ .

5. How many hours have passed?

_____ hours have passed.

6. How many hours have passed?

_____ hours have passed.

7. Darla was in line at 4:00. She waited 15 minutes to get on the ride. What time did she get on the ride?

8. Derek had a dentist appointment at 3:10. He left 20 minutes later. What time did Derek leave the dentist office?

Practice 31

Directions: Use this calendar to answer the questions below.

February

Sunday	Monday	Tuesday	Wednesday	Thursday	Friday	Saturday
			1	2	3	4
5	6	7	8	9	10	11
12	13	14	15	16	17	18
19	20	21	22	23	24	25
26	27	28	29			

1. Matthew plans to go to the movies with Douglas on February 4.
   What day of the week are they going to the movies? _____

2. Katrina has a library club meeting on the third Thursday of February.
   What is the date of her meeting? _____

3. Samantha has gymnastics classes every Tuesday.
   How many classes will she have during February? _____

4. E. J. plans to spend a week at his grandma's. If he arrives on February 13,
   on what day of the week and on what date will he leave his grandma's
   house?

   _____

5. Dylan has a basketball game to play on the last Wednesday
   of the month. What is the date of his basketball game? _____

Practice 32

1. What day of the week is March 14?

MARCH						
SUN	MON	TUE	WED	THU	FRI	SAT
1	2	3	4	5	6	7
8	9	10	11	12	13	14
15	16	17	18	19	20	21
22	23	24	25	26	27	28
29	30	31				

(A) Wednesday   (B) Thursday   (C) Saturday   (D) Monday

---

2. What day of the week is September 17?

SEPTEMBER						
SUN	MON	TUE	WED	THU	FRI	SAT
			1	2	3	4
5	6	7	8	9	10	11
12	13	14	15	16	17	18
19	20	21	22	23	24	25
26	27	28	29	30		

(A) Tuesday   (B) Saturday   (C) Sunday   (D) Friday

Dates and Calendars

Practice 33

1.

	September					
Sunday	Monday	Tuesday	Wednesday	Thursday	Friday	Saturday
	1	2	3	4	5	6
7	8	9	10	11	12	13
14	15	16	17	18	19	20
21	22	23	24	25	26	27
28	29	30				

Which of these dates is on a Wednesday?
(A) September 11   (B) September 3
(C) September 19   (D) September 16

2.

	October					
Sunday	Monday	Tuesday	Wednesday	Thursday	Friday	Saturday
	1	2	3	4	5	6
7	8	9	10	11	12	13
14	15	16	17	18	19	20
21	22	23	24	25	26	27
28	29	30	31			

Which of these dates is on a Sunday?
(A) October 3   (B) October 28
(C) October 2   (D) October 13

Practice 34

1. Look at the two clocks.

   The second clock has a later time than the first clock.

   How many hours later is the second clock than the first clock?

   (A) 3 hours later   (B) 2 hours later

   (C) 4 hours later   (D) 1 hour later

---

2. Look at the two clocks.

   The second clock has a later time than the first clock.

   How many hours later is the second clock than the first clock?

   (A) 3 hours later   (B) 1 hour later

   (C) 4 hours later   (D) 2 hours later

Practice 35

1. Look at the two clocks.

   The second clock has a later time than the first clock.

   How many hours later is the second clock than the first clock?

(A) 4 hours later   (B) 1 hour later

(C) 3 hours later   (D) 2 hours later

---

2. Look at the two clocks.

   The second clock has a later time than the first clock.

   How many hours later is the second clock than the first clock?

(A) 3 hours later   (B) 4 hours later

(C) 1 hour later   (D) 2 hours later

Practice 36

Directions: Read each statement. Circle A.M. or P.M. to show the time of day for each event.

1. Jonas eats breakfast at 7:00 in the morning.

   (A.M.)   P.M.

2. Gwen does her homework every afternoon at 4:00.

   A.M.   (P.M.)

3. Ramon sets the table for dinner every evening at 5:00.

   A.M.   (P.M.)

4. Sabrina arrives home from school at 3:00 each afternoon.

   A.M.   (P.M.)

5. Paolo plays in the park at 2:00 in the afternoon.

   A.M.   (P.M.)

6. Eva watches the Saturday morning cartoons at 9:00.

   (A.M.)   P.M.

7. Cedric takes his bath each night at 8:00.

   A.M.   (P.M.)

8. Abby wakes up at 6:00 each morning.

   (A.M.)   P.M.

Test Practice 1

For each problem determine what time each clock shows.

1.

(A) 7:00

(B) 6:00

(C) 5:00

(D) 4:00

2.

(A) 1:00

(B) 11:00

(C) 12:00

(D) 10:00

3.

(A) 8:00

(B) 9:00

(C) 10:00

(D) 11:00

4.

(A) 3:00

(B) 4:00

(C) 5:00

(D) 6:00

5.

(A) 6:00

(B) 7:00

(C) 9:00

(D) 8:00

6.

(A) 12:00

(B) 1:30

(C) 11:00

(D) 2:30

7.

(A) 6:30

(B) 8:30

(C) 9:30

(D) 7:30

8.

(A) 4:30

(B) 12:30

(C) 6:30

(D) 10:30

9.

(A) 10:30

(B) 7:30

(C) 4:30

(D) 9:30

10.

(A) 4:30

(B) 3:30

(C) 5:30

(D) 6:30

Test Practice 2

For each problem, determine what time each clock shows.

1.

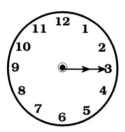

- (A) 4:15
- (B) 3:15
- (C) 6:15
- (D) 5:15

2.

- (A) 5:45
- (B) 4:45
- (C) 6:45
- (D) 7:45

3.

- (A) 11:15
- (B) 12:15
- (C) 11:20
- (D) 12:20

4.

- (A) 6:30
- (B) 6:45
- (C) 6:15
- (D) 6:00

5.

- (A) 8:50
- (B) 9:00
- (C) 8:45
- (D) 9:15

6.

- (A) 12:00
- (B) 12:05
- (C) 12:10
- (D) 12:15

7.

- (A) 10:00
- (B) 10:15
- (C) 10:30
- (D) 10:40

8.

- (A) 6:00
- (B) 6:20
- (C) 6:40
- (D) 6:30

9.

- (A) 8:15
- (B) 8:20
- (C) 8:10
- (D) 8:25

10.

- (A) 10:50
- (B) 10:40
- (C) 11:00
- (D) 11:10

Test Practice 3

For each problem determine what time each clock shows.

1.

(A) 4:20

(B) 4:45

(C) 4:30

(D) 4:35

2.

(A) 5:45

(B) 5:50

(C) 5:55

(D) 6:00

3.

(A) 9:40

(B) 9:30

(C) 9:50

(D) 9:45

4.

(A) 8:00

(B) 8:10

(C) 8:05

(D) 8:15

5.

(A) 10:30

(B) 10:40

(C) 10:45

(D) 10:35

6.

(A) 11:20

(B) 11:30

(C) 11:25

(D) 11:35

7.

(A) 8:40

(B) 8:30

(C) 8:50

(D) 8:20

8.

(A) 1:40

(B) 2:40

(C) 1:45

(D) 2:45

9.

(A) 10:00

(B) 10:10

(C) 10:15

(D) 10:05

10.

(A) 8:00

(B) 7:55

(C) 7:50

(D) 8:05

Test Practice 4

Determine how much time has passed between each set of clocks.

1.

Ⓐ 2 hours
Ⓑ 3 hours
Ⓒ 4 hours
Ⓓ 5 hours

6.

Ⓐ 4 hours
Ⓑ 2 hours
Ⓒ 3 hours
Ⓓ 5 hours

2.

Ⓐ 5 hours
Ⓑ 4 hours
Ⓒ 3 hours
Ⓓ 2 hours

7.

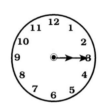 

Ⓐ 3 hours
Ⓑ 4 hours
Ⓒ 2 hours
Ⓓ 1 hour

3.

Ⓐ 5 hours
Ⓑ 4 hours
Ⓒ 6 hours
Ⓓ 3 hours

8.

Ⓐ 6 hours
Ⓑ 1 hour
Ⓒ 2 hours
Ⓓ 3 hours

4.

Ⓐ 4 hours
Ⓑ 3 hours
Ⓒ 5 hours
Ⓓ 6 hours

9.

Ⓐ 6 hours
Ⓑ 7 hours
Ⓒ 10 hours
Ⓓ 12 hours

5.

Ⓐ 5 hours
Ⓑ 4 hours
Ⓒ 3 hours
Ⓓ 2 hours

10.

Ⓐ 7 hours
Ⓑ 8 hours
Ⓒ 6 hours
Ⓓ 5 hours

Test Practice 5 ◌ ◌ ◌ ◌ ◌ ◌ ◌ ◌ ◌ ◌ ◌ ◌ ◌

Answer each question.

1.

1 minute = ___ seconds

(A) 60
(B) 50
(C) 40
(D) 30

2.

1 hour = ___ minutes

(A) 30
(B) 60
(C) 20
(D) 100

3.

1 day = ___ hours

(A) 36
(B) 48
(C) 24
(D) 12

4.

1 week = ___ days

(A) 10
(B) 7
(C) 5
(D) 4

5.

1 decade = ___ years

(A) 10
(B) 12
(C) 6
(D) 9

6.

What day of the week follows Monday?

(A) Tuesday
(B) Wednesday
(C) Saturday
(D) Sunday

7.

What day of the week follows Tuesday?

(A) Thursday
(B) Friday
(C) Wednesday
(D) Monday

8.

What day of the week follows Friday?

(A) Tuesday
(B) Saturday
(C) Friday
(D) Sunday

9.

What day of the week follows Saturday?

(A) Sunday
(B) Tuesday
(C) Monday
(D) Wednesday

10.

What day of the week follows Thursday?

(A) Friday
(B) Wednesday
(C) Thursday
(D) Saturday

Test Practice 6

Answer each question.

1.

What month comes after January?

Ⓐ February
Ⓑ March
Ⓒ April
Ⓓ May

2.

What month comes after September?

Ⓐ May
Ⓑ June
Ⓒ October
Ⓓ November

3.

What month comes before November?

Ⓐ December
Ⓑ October
Ⓒ September
Ⓓ August

4.

What month comes before July?

Ⓐ May
Ⓑ August
Ⓒ June
Ⓓ September

5.

What month is between March and May?

Ⓐ April
Ⓑ February
Ⓒ June
Ⓓ July

6.

What month is between July and September?

Ⓐ August
Ⓑ October
Ⓒ June
Ⓓ May

7.

What is the 7th month in a calendar year?

Ⓐ July
Ⓑ June
Ⓒ May
Ⓓ August

8.

What is the 12th month in a calendar year?

Ⓐ December
Ⓑ October
Ⓒ November
Ⓓ September

9.

What month has only 28 days (or 29 days in a leap year)?

Ⓐ April
Ⓑ March
Ⓒ January
Ⓓ February

10.

There are _____ months in a year.

Ⓐ 11
Ⓑ 10
Ⓒ 12
Ⓓ 9

Answer Sheet

Test Practice 1	Test Practice 2	Test Practice 3
1. Ⓐ Ⓑ Ⓒ Ⓓ	1. Ⓐ Ⓑ Ⓒ Ⓓ	1. Ⓐ Ⓑ Ⓒ Ⓓ
2. Ⓐ Ⓑ Ⓒ Ⓓ	2. Ⓐ Ⓑ Ⓒ Ⓓ	2. Ⓐ Ⓑ Ⓒ Ⓓ
3. Ⓐ Ⓑ Ⓒ Ⓓ	3. Ⓐ Ⓑ Ⓒ Ⓓ	3. Ⓐ Ⓑ Ⓒ Ⓓ
4. Ⓐ Ⓑ Ⓒ Ⓓ	4. Ⓐ Ⓑ Ⓒ Ⓓ	4. Ⓐ Ⓑ Ⓒ Ⓓ
5. Ⓐ Ⓑ Ⓒ Ⓓ	5. Ⓐ Ⓑ Ⓒ Ⓓ	5. Ⓐ Ⓑ Ⓒ Ⓓ
6. Ⓐ Ⓑ Ⓒ Ⓓ	6. Ⓐ Ⓑ Ⓒ Ⓓ	6. Ⓐ Ⓑ Ⓒ Ⓓ
7. Ⓐ Ⓑ Ⓒ Ⓓ	7. Ⓐ Ⓑ Ⓒ Ⓓ	7. Ⓐ Ⓑ Ⓒ Ⓓ
8. Ⓐ Ⓑ Ⓒ Ⓓ	8. Ⓐ Ⓑ Ⓒ Ⓓ	8. Ⓐ Ⓑ Ⓒ Ⓓ
9. Ⓐ Ⓑ Ⓒ Ⓓ	9. Ⓐ Ⓑ Ⓒ Ⓓ	9. Ⓐ Ⓑ Ⓒ Ⓓ
10. Ⓐ Ⓑ Ⓒ Ⓓ	10. Ⓐ Ⓑ Ⓒ Ⓓ	10. Ⓐ Ⓑ Ⓒ Ⓓ

Test Practice 4	Test Practice 5	Test Practice 6
1. Ⓐ Ⓑ Ⓒ Ⓓ	1. Ⓐ Ⓑ Ⓒ Ⓓ	1. Ⓐ Ⓑ Ⓒ Ⓓ
2. Ⓐ Ⓑ Ⓒ Ⓓ	2. Ⓐ Ⓑ Ⓒ Ⓓ	2. Ⓐ Ⓑ Ⓒ Ⓓ
3. Ⓐ Ⓑ Ⓒ Ⓓ	3. Ⓐ Ⓑ Ⓒ Ⓓ	3. Ⓐ Ⓑ Ⓒ Ⓓ
4. Ⓐ Ⓑ Ⓒ Ⓓ	4. Ⓐ Ⓑ Ⓒ Ⓓ	4. Ⓐ Ⓑ Ⓒ Ⓓ
5. Ⓐ Ⓑ Ⓒ Ⓓ	5. Ⓐ Ⓑ Ⓒ Ⓓ	5. Ⓐ Ⓑ Ⓒ Ⓓ
6. Ⓐ Ⓑ Ⓒ Ⓓ	6. Ⓐ Ⓑ Ⓒ Ⓓ	6. Ⓐ Ⓑ Ⓒ Ⓓ
7. Ⓐ Ⓑ Ⓒ Ⓓ	7. Ⓐ Ⓑ Ⓒ Ⓓ	7. Ⓐ Ⓑ Ⓒ Ⓓ
8. Ⓐ Ⓑ Ⓒ Ⓓ	8. Ⓐ Ⓑ Ⓒ Ⓓ	8. Ⓐ Ⓑ Ⓒ Ⓓ
9. Ⓐ Ⓑ Ⓒ Ⓓ	9. Ⓐ Ⓑ Ⓒ Ⓓ	9. Ⓐ Ⓑ Ⓒ Ⓓ
10. Ⓐ Ⓑ Ⓒ Ⓓ	10. Ⓐ Ⓑ Ⓒ Ⓓ	10. Ⓐ Ⓑ Ⓒ Ⓓ

Answer Key

Page 4
1. 1:00
2. 9:00
3. 6:00
4. 8:00
5. 11:00
6. 4:00
7. 3:00
8. 12:00
9. 5:00
10. 10:00
11. 2:00
12. 7:00

Page 5
1. A
2. C

Page 6
1. D
2. B

Page 7
1. A
2. C

Page 8
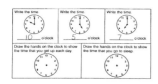

Page 9
1. 5:00
2. 4:00
3. 9:00
4. 7:00
5. 2:00
6. 6:00

Page 10
1. 1:00
2. 8:00
3. 12:00
4. 10:00
5. 4:00
6. 11:00

Page 11
1. D
2. A

Page 12
1. A
2. C

Page 13
1. 5
2. 7
3. 5
4. 4
5. 8
6. 3

Page 14
1. 6
2. 1
3. 2
4. 10
5. 3
6. 10

Page 15
1. 12:30
2. 7:30
3. 8:30
4. 4:30
5. 9:30
6. 10:30

Page 16
1. C
2. B

3. D
4. C

Page 17
1. B
2. B
3. A
4. D

Page 18
1.
4.
2.
5.
3.
6.

Page 19
1. B
2. B
3. C
4. C

Page 20
1. B
2. A
3. D
4. B

Page 21
1. B
2. D
3. D
4. A

Page 22
1. A
2. C
3. A
4. A

Page 23
1. B
2. C
3. A
4. C

Page 24
1. 4.
2.  5.
3. 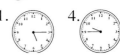 6.

Page 25
1.  4.
2. 5.
3. 6.

Page 26
1. 9:35
2. 10:50
3. 5:15
4. 3:25
5. 12:10
6. 2:40

Page 27
1. 9:55
2. 9:10
3. 5:40
4. 3:05
5. 10:20
6. 2:20

Answer Key

Page 28

1. 6:10
2. 1:50
3. 1:40
4. 4:10
5. 2:40
6. 7:20

Page 29

1. 2:50
2. 1:20
3. 11:25
4. 9:20
5. 3:30
6. 10:25

Page 30

1. 2:30
2. 3:00
3. 3:30
4. 4:00; 4:30

Page 31

1. 4:00 P.M.–
   6:00 P.M.
2. 15 books
3. 8:00 A.M.–
   10:00 A.M.
4. 10:00 A.M.–
   12:00 P.M. and
   12:00 P.M.–
   2:00 P.M.
5. 25 books

Page 32

1. 6:45 P.M.
2. 12:15 P.M.
3. 1 hour
   15 minutes
4. 4 P.M.
5. 2:30 P.M.
6. 8:00 P.M.
7. 2 hours
   30 minutes

Page 33

1. P.M.
2. A.M.
3. A.M.
4. P.M.
5. 2
6. 3
7. 4:15
8. 3:30

Page 34

1. Saturday
2. February 16
3. 4 classes
4. February 20,
   Monday
5. February 29

Page 35

1. C
2. D

Page 36

1. B
2. B

Page 37

1. B
2. D

Page 38

1. D
2. D

Page 39

1. A.M.
2. P.M.
3. P.M.
4. P.M.
5. P.M.
6. A.M.
7. P.M.
8. A.M.

Page 40

1. A
2. C
3. B
4. C
5. A
6. B
7. D
8. A
9. A
10. B

Page 41

1. B
2. B
3. B
4. C
5. C
6. C
7. D
8. D
9. B
10. A

Page 42

1. B
2. C
3. D
4. C
5. D
6. C
7. C
8. B
9. B
10. B

Page 43

1. B
2. D
3. C
4. A
5. B
6. C
7. A
8. D
9. A
10. A

Page 44

1. A
2. B
3. C
4. B
5. A
6. A
7. C
8. B
9. A
10. A

Page 45

1. A
2. C
3. B
4. C
5. A
6. A
7. A
8. A
9. D
10. C

#3317 Practice Makes Perfect: Time